RHYTHM TO THE UNIVERSE

Why am doing this mini-journal?

I never believed that the calendar started in the middle of winter. It seemed so un-natural to me as a child. Growing up on a farm, I witnessed life beginning in the spring; the vegetation starts blooming and the baby farm and wild animals are born all over creation. And, I re-discovered, after a lifelong question; that the calendar does begin in the spring.

What is a lunar-solar calendar and what can it do for you?

You will be in tune to the rhythm of the Universe.

God originally gave this calendar to the 12 Tribes of Israel, known today as the Jews.

—To Judah- the only tribe we recognize

Also, credit given:

Dr. David H. Stern, the Complete Jewish Bible, and study Bible, also HolyLanguage.com.

Rhythm to the Universe
Moon means month!
GW Kushner

Simplify this by looking at a clock. You will see it is divided into 12 equal parts. If you draw a line from the 12 to the 6 we can call this the equinox.

In the Northern Hemisphere, above the equator, imagine our spring starts at the number 6. Our fall starts roughly at the number 12. If you live below the equator these seasons are opposite.

To explain the starting of the lunar – solar calendar:

The numerical order for the months begins in the springtime, at the number 6 on the clock, the 1st Moon month.

The number 12 on the clock signals the new year beginning – this is the 7th Moon month.

The calendar we have known begins the year and the calendar on the number 3 of the clock, January.

The First Moon month heralds the springtime and the beginning of the calendar year. Moon months are numbered from this starting point. This year we had a 13th Moon. This happens 7 times out of a 19- year cycle.

The Seventh Moon is "Rosh Hodesh." This is the language of Hebrew; the original language and "Torah" is translated as teaching; this is what is known today as The Holy Bible. Rosh Hodesh literally is the "head of the year". We comprehend this as the New Year. "The turn of the year," to be poetic, I found in one version of the Bible.

Every Moon month is twenty-nine and a half days long. So, a Moon month will have either 29- or 30-day cycle.

The Moon is new when the month begins, and it is full in the middle of the month on the 14, 15 and 16th.

The calendar does begin in spring, like I thought it should, when I was a child.

The civil year begins in the fall. This is documented in the Holy Bible, recorded in Genesis 8:4; the new year was commemorated in ancient history when the flood waters brought the ark to a rest on the top of the highest mountain.
Spring is when the flood waters were receded and Noah, his family and the animals were able to climb out of the ark and start down the mountain. And spring also has to do with when they boarded the ark. You can read about this in Genesis chapters eight and nine.

In the book of Exodus, the calendar sets forth Passover, Firstfruits and Sukkot; the Holy Days we are instructed to keep, originally. But, because I was brought up in a Christian household, I found the rhythm for the 12 and 13 months in the dates of Easter. You will see it too, if you look over a 19 year or more span. The dates coincide; which tells me that the Christian dates of Easter are based on the calendar Our Father set forth in the books of Genesis and Exodus.

The new Moon as a start for the month is mentioned throughout the Word of God, for the new Moon was another day held in reverence, to honor God. In Psalm 104:19 King David notices, "you made the Moon to mark the seasons and the Sun knows when to set."

The following is a song I composed about this topic....

It is also the title to this mini journal.

RHYTHM TO THE UNIVERSE

There's a rhythm to the Universe that we don't understand.
It existed since the beginning, long before the start of man.
I'd love to share this with you, but only if you can understand, understand.

I'm going to tell you all a story about how it all began,
The story of creation, the Heavens and the Earth.
God set the Sun and Moon in place and the Universe gave birth,
To give light to the Earth, to give light to the Earth.
It was on the fourth day of creation, before the start of man,
God placed the Moon to guide our months and the Sun to rule the year.
He placed two lights in the dome, to give light to the Earth, to give light to the Earth.

The words make such a stage for it, but the music makes it real.
You've got to get the rhythm that the music makes you feel.
The Universe in harmony, interval our tonality,
But we are not in tune, no we are not in tune.

There's a rhythm to the Universe that we don't understand.
It existed since the beginning, long before the start of man.
I'd love to share this with you, but only if you can, understand, please understand.

Everything on Earth is guided by the rhythms of the Moon,
The seasons and Earth's produce, the animals and our cycles, the ocean and its fishes, but not man, but not man.

All creation knows this rhythm that isn't known by man,
   to be in sync with is to be all that we can.
   People won't remember what you said or did in June.
   All they will remember is if you are in tune.

   There are two lights in the sky, the Moon is not to fear.
   The Sun rules the day and the length of the year.
   What we are missing is the rhythms of the Moon.
   We need to get this rhythm, so we can be in tune.

(This song was created with my friend and mentor, Israeli Rothman in 2014) GW Kushner

RHYTHM TO THE UNIVERSE

In the Lunar Solar Calendar, Moon means month!

"Every month on Rosh Hodesh (new Moon)
And every week on Shabbat, (Sabbath)
Everyone living will come
To worship in my presence,"
 Says ADONAI
—Isaiah 66:23

How to get in tune with the rhythm of the Universe?

Just imagine you are back in time, long before modern civilization – before automobiles, television, telephone or computers. How did we measure time?

Moon was the original meaning of the word month, and that is how we kept track of time. God gave the Hebrew children the knowledge of the workings of Sun and Moon with Earth.

It took the rest of civilization until the 1500's when Lilius fine-tuned it. (Aloysius Lilius 1510-1576 was an Italian doctor, astronomer, philosopher and chronologist.) Pope Gregory XIII took credit after Lilius' death with the Gregorian calendar. His Holiness used Lilius' findings, and did not credit him, so the astronomers named the crater on the Moon after him.

By the 1500's the spring equinox was off schedule by 10 days, so jump starting it ten days and taking out some leap years in February's schedule every four years, fixed the timing.

But this was not added to the decree; His Holiness secretly added this in: changing the beginning of the calendar from spring to the fixed winter solstice.

The Almighty, in his infinite wisdom has had this calendar in place all along. It has never changed, or the seventh day of the week. The weeks and the days are ruled by the Sun. The months are ruled by the Moon.

Psalm 104:19 "You made the Moon to mark the seasons, and the Sun knows when to set."

The First Moon starts in the Spring of the year. The turn of the year, or the new year is in the Fall, the 7th Moon.

Ordered creation calls for a Creator.

This is proof that we have a Creator because the 12 Hebrew tribes already knew this. The Sun ruled the year, with the 7th day Sabbaths being in accordance with the Sun.

The Moon ruled the Holy Days set forth by God; the beginning of the calendar, the beginning of the year, and more. There are 4 heads to the year in Hebrew calendar keeping.

These Holy Days are teachings in the Ten Commandments:

Passover, first fruits (or Pentecost,) and Sukkot (the feast of ingathering). The early Christians kept God's original Holy Days, but they were persecuted by the Roman Catholic Church, who persisted in their man-made holy days that included Easter and Christmas, be mandated. Now, if you don't celebrate these traditions, it is perceived that you don't believe in God. God's Holy Days were corrupted, and a lie became the truth, and the truth was persecuted and flung to the ground.

Christmas is based on the yearly cycle of the Sun, disguised by the supposed birth of Christ, it is no secret now; instituted to be a worship of Ba'al; December 25. And Easter is a story to secretly worship Ishtar, the Babylonian fertility goddess.

But the remains of this mysterious lunar calendar are present in the dates of Easter. Easter is based on the lunar-solar calendar.

The dates of Easter differ from year to year. You will see the date fall back 11 days, and another 11 days and then leap ahead the following year by 21 days. This is because the Moon cycle is about 11 days shorter than the cycle of the Sun per year.

So, in the 19-year cycle, 12 years have 12 Moon months and 7 years have an extra month, known as 2nd Adar in the Hebrew months. God gave this calendar and thereby; gave Sun and Moon to all nations and told us not to worship these luminaries.

I just give the number of the months, not the Jewish name for the months in this min-journal. This is known as a "Jewish calendar," because the tribe of Judah is the only tribe left who kept God's original Holy Days, designated by a lunar -solar calendar. This calendar contains the mysterious extra Moon.

GW Kushner

The
Journal

The journal consists of a month and blank pages with dividers for the days of the week with the calendar as the divider of the Natural Moon months. As a daily journal I gave different ideas on how to use this journal:

Write the Gregorian Date: first and the Moon Date: second, and after a while you will know what phase the Moon was in by the date you write it. Leaving one page, per day was too much, so I have provided a page a week. You will be able to section off the week as you make an entry for each day.

(for example) March 1, 2019 and underneath it- write the "12 Moon 26th day". This tells you that the Moon month is almost over.

Doing this gets you in tune with the cycles of the Moon. Then, make a journal entry or–to visualize a goal, write corresponding words to claim it (faith) to act like this has already happened.

Even Jesus spoke about this power when questioned about adultery, he said something like, even if you think of doing (thought plus emotion) it you have already done it. That is the re-uniting of your emotion with your thought.

We don't know that we have this power. And we won't know if we don't put it to use. Didn't Jesus say, time and time again, "…by your faith you are healed." He said, "Go, and sin no more," and, "You shall do these things and greater?"

A gratitude journal – take note of things you are grateful for – this creates a feeling of well-being and satisfaction.

And, look at the date and months spoken about in Biblical History, because these are the corresponding months.

Remember, this is not a chore, and it's ok if it turns into a to-do list because to achieve your goals, you act, you take calculated risks and you have a list of things to do that you do not want to forget. As a way of keeping organized: are you able to schedule in time to relax, be in the moment and cultivate a feeling of gratitude and happiness?

Remember the Sabbath day, to keep it holy. This is the day that means seven, and it is Saturday.

In my opinion, it was changed to Sunday maybe because Yesuha (Jesus) first appeared to them on the going out of Shabbat, which is Saturday evening, misconstrued as Sunday.

The Sabbath is a sign between God and his people, and God did not change it to Sunday. It has been the same throughout ancient and modern history.

RHYTHM TO THE UNIVERSE

A new Moon is for introversion, for planning as well as resting along with the last few days waning, beforehand. Starting something new, you would want to start on a new Moon, an opening, for example of a restaurant or any business, it's a brand-new beginning.

The full Moon (15, 16 and 17) is what you shoot for to complete a goal, and you have the rest of the month to finish it, if you didn't. After the full Moon, you would "wrap things up." It is a time for reflection.

You could even draw the phases of the Moon in at first. If you want to get rid of bad habits, break them by starting good habits in a new Moon beginning.

After the full Moon, is a good time for shedding unwanted things, even weight. This is the give-away time.
Donations to charity?

If planting, depending on the seed, it takes about a ten days week for it to take root and sprout if it is germinated already, it will pop out of the Earth on a new Moon!

Your Journal is your key to a greater awareness of the makings of the Universe, given to us by the Eternal, Universal God.

Here are my usage points:

Use the Moon date and the Gregorian date. You might want to draw the Moon phases on your Moon month until you are accustomed to it. Every Moon month begins with a new Moon -dark- introversion and reflection and planning. This would be a good discipline to chart your journal in the last few days before the new Moon month.

There may be a few extra days in the weekly calendar, left over from the month before, use these spaces to plan. Five or six weeks will be given for every month to accommodate for this, and because the months don't always begin on a Sunday.

RHYTHM TO THE UNIVERSE

2019: 13th Moon (March 6 - April 4)

Sun	Mon	Tues	Wed	Thurs	Fri	Sat
			March 6 1	March 7 2	March 8 3	March 9 4
March 10 5	March 11 6	March 12 7	March 13 8	March 14 9	March 15 10	March 16 11
March 17 12	March 18 13	March 19 14	March 20 15	March 21 16	March 22 17	March 23 18
March 24 19	March 25 20	March 26 21	March 27 22	March 28 23	March 29 24	March 30 25
March 31 26	April 1 27	April 2 28	April 3 29	April 4 30		

Sunday:
Monday:
Tuesday:
Wednesday:
Thursday:
Friday:
Saturday:

Sunday:	
Monday:	
Tuesday:	
Wednesday:	
Thursday:	
Friday:	
Saturday:	

Sunday:
Monday:
Tuesday:
Wednesday:
Thursday:
Friday:

Saturday:

Sunday:

Monday:

Tuesday:

Wednesday:

Thursday:

Friday:
Saturday:

Sunday:
Monday:
Tuesday:
Wednesday:
Thursday:
Friday:
Saturday:

2019: 1st Moon (April 5 - May 3)

Sun	Mon	Tues	Wed	Thurs	Fri	Sat
					April 5 1	April 6 2
April 7 3	April 8 4	April 9 5	April 10 6	April 11 7	April 12 8	April 13 9
April 14 10	April 15 11	April 16 12	April 17 13	April 18 14	April 19 15	April 20 16
April 21 17	April 22 18	April 23 19	April 24 20	April 25 21	April 26 22	April 27 23
April 28 24	April 29 25	April 30 26	May 1 27	May 2 28	May 3 29	

Sunday:

Monday:

Tuesday:

Wednesday:

Thursday:

Friday:

Saturday:

Sunday:
Monday:
Tuesday:
Wednesday:
Thursday:
Friday:
Saturday:

RHYTHM TO THE UNIVERSE

Sunday:

Monday:

Tuesday:

Wednesday:

Thursday:

Friday:

aturday:

Sunday:
Monday:
Tuesday:
Wednesday:
Thursday:
Friday:

Saturday:

Sunday:
Monday:
Tuesday:
Wednesday:
Thursday:
Friday:
Saturday:

2019: 2nd Moon (May 4 - June 2)

Sun	Mon	Tues	Wed	Thurs	Fri	Sat
						May 4 **1**
May 5 **2**	May 6 **3**	May 7 **4**	May 8 **5**	May 9 **6**	May 10 **7**	May 11 **8**
May 12 **9**	May 13 **10**	May 14 **11**	May 15 **12**	May 16 **13**	May 17 **14**	May 18 **15**
May 19 **16**	May 20 **17**	May 21 **18**	May 22 **19**	May 23 **20**	May 24 **21**	May 25 **22**
May 26 **23**	May 27 **24**	May 28 **25**	May 29 **26**	May 30 **27**	May 31 **28**	June 1 **29**
June 2 **30**						

Sunday:
Monday:
Tuesday:
Wednesday:
Thursday:
Friday:
Saturday:

Sunday:
Monday:
Tuesday:
Wednesday:
Thursday:
Friday:
Saturday:

Sunday:

Monday:

Tuesday:

Wednesday:

Thursday:

Friday:

Saturday:

Sunday:

Monday:

Tuesday:

Wednesday:

Thursday:

Friday:
Saturday:

RHYTHM TO THE UNIVERSE

Sunday:

Monday:

Tuesday:

Wednesday:

Thursday:

Friday:

Saturday:

2019: 3rd Moon (June 3 - July 1)

Sun	Mon	Tues	Wed	Thurs	Fri	Sat
	June 3 1	June 4 2	June 5 3	June 6 4	June 7 5	June 8 6
June 9 7	June 10 8	June 11 9	June 12 10	June 13 11	June 14 12	June 15 13
June 16 14	June 17 15	June 18 16	June 19 17	June 20 18	June 21 19	June 22 20
June 23 21	June 24 22	June 25 23	June 26 24	June 27 25	June 28 26	June 29 27
June 30 28	July 1 29					

Sunday:

Monday:

Tuesday:

Wednesday:

Thursday:

Friday:

Saturday:

Sunday:
Monday:
Tuesday:
Wednesday:
Thursday:
Friday:
Saturday:

RHYTHM TO THE UNIVERSE

Sunday:

Monday:

Tuesday:

Wednesday:

Thursday:

Friday:

Saturday:

Sunday:
Monday:
Tuesday:
Wednesday:
Thursday:
Friday:

RHYTHM TO THE UNIVERSE

Saturday:

Sunday:

Monday:

Tuesday:

Wednesday:

Thursday:

Friday:

Saturday:

2019: 4th Moon (July 2 - July 31)

Sun	Mon	Tues	Wed	Thurs	Fri	Sat
		July 2 1	July 3 2	July 4 3	July 5 4	July 6 5
July 7 6	July 8 7	July 9 8	July 10 9	July 11 10	July 12 11	July 13 12
July 14 13	July 15 14	July 16 15	July 17 16	July 18 17	July 19 18	July 20 19
July 21 20	July 22 21	July 23 22	July 24 23	July 25 24	July 26 25	July 27 26
July 28 27	July 29 28	July 30 29	July 31 30			

Sunday:

Monday:

Tuesday:

Wednesday:

Thursday:

Friday:

Saturday:

Sunday:
Monday:
Tuesday:
Wednesday:
Thursday:
Friday:
Saturday:

Sunday:

Monday:

Tuesday:

Wednesday:

Thursday:

Friday:

Saturday:

Sunday:

Monday:

Tuesday:

Wednesday:

Thursday:

Friday:

Saturday:

RHYTHM TO THE UNIVERSE

Sunday:

Monday:

Tuesday:

Wednesday:

Thursday:

Friday:

Saturday:

2019: 5th Moon (Aug 1 - Aug 29)

Sun	Mon	Tues	Wed	Thurs	Fri	Sat
				Aug 1 1	Aug 2 2	Aug 3 3
Aug 4 4	Aug 5 5	Aug 6 6	Aug 7 7	Aug 8 8	Aug 9 9	Aug 10 10
Aug 11 11	Aug 12 12	Aug 13 13	Aug 14 14	Aug 15 15	Aug 16 16	Aug 17 17
Aug 18 18	Aug 19 19	Aug 20 20	Aug 21 21	Aug 22 22	Aug 23 23	Aug 24 24
Aug 25 25	Aug 26 26	Aug 27 27	Aug 28 28	Aug 29 29		

Sunday:	
Monday:	
Tuesday:	
Wednesday:	
Thursday:	
Friday:	
Saturday:	

Sunday:

Monday:

Tuesday:

Wednesday:

Thursday:

Friday:

Saturday:

Sunday:
Monday:
Tuesday:
Wednesday:
Thursday:
Friday:
aturday:

Sunday:
Monday:
Tuesday:
Wednesday:
Thursday:
Friday:

Saturday:

Sunday:
Monday:
Tuesday:
Wednesday:
Thursday:
Friday:
Saturday:

RHYTHM TO THE UNIVERSE

2019: 6th Moon (Aug 30 - Sept 27)

Sun	Mon	Tues	Wed	Thurs	Fri	Sat
					Aug 30 **1**	Aug 31 **2**
Sept 1 **3**	Sept 2 **4**	Sept 3 **5**	Sept 4 **6**	Sept 5 **7**	Sept 6 **8**	Sept 7 **9**
Sept 8 **10**	Sept 9 **11**	Sept 10 **12**	Sept 11 **13**	Sept 12 **14**	Sept 13 **15**	Sept 14 **16**
Sept 15 **17**	Sept 16 **18**	Sept 17 **19**	Sept 18 **20**	Sept 19 **21**	Sept 20 **22**	Sept 21 **23**
Sept 22 **24**	Sept 23 **25**	Sept 24 **26**	Sept 25 **27**	Sept 26 **28**	Sept 27 **29**	

Sunday:

Monday:

Tuesday:

Wednesday:

Thursday:

Friday:

Saturday:

Sunday:

Monday:

Tuesday:

Wednesday:

Thursday:

Friday:

Saturday:

Sunday:

Monday:

Tuesday:

Wednesday:

Thursday:

Friday:

Saturday:

Sunday:

Monday:

Tuesday:

Wednesday:

Thursday:

Friday:
Saturday:

Sunday:
Monday:
Tuesday:
Wednesday:
Thursday:
Friday:
Saturday:

2019: 7th Moon (Sept 28 - Oct 26)

Sun	Mon	Tues	Wed	Thurs	Fri	Sat
			The Turn of the Year			Sept 28 1
Sept 29 2	Sept 30 3	Oct 1 4	Oct 2 5	Oct 3 6	Oct 4 7	Oct 5 8
Oct 6 9	Oct 7 10	Oct 8 11	Oct 9 12	Oct 10 13	Oct 11 14	Oct 12 15
Oct 13 16	Oct 14 17	Oct 15 18	Oct 16 19	Oct 17 20	Oct 18 21	Oct 19 22
Oct 20 23	Oct 21 24	Oct 22 25	Oct 23 26	Oct 24 27	Oct 25 28	Oct 26 29

Sunday:

Monday:

Tuesday:

Wednesday:

Thursday:

Friday:

Saturday:

Sunday:
Monday:
Tuesday:
Wednesday:
Thursday:
Friday:
Saturday:

Sunday:	
Monday:	
Tuesday:	
Wednesday:	
Thursday:	
Friday:	
Saturday:	

| Sunday: |
| Monday: |
| Tuesday: |
| Wednesday: |
| Thursday: |
| Friday: |

Saturday:

Sunday:
Monday:
Tuesday:
Wednesday:
Thursday:
Friday:
Saturday:

2019: 8th Moon (Oct 27 - Nov 25)

Sun	Mon	Tues	Wed	Thurs	Fri	Sat
Oct 27 **1**	Oct 28 **2**	Oct 29 **3**	Oct 30 **4**	Oct 31 **5**	Nov 1 **6**	Nov 2 **7**
Nov 3 **8**	Nov 4 **9**	Nov 5 **10**	Nov 6 **11**	Nov 7 **12**	Nov 8 **13**	Nov 9 **14**
Nov 10 **15**	Nov 11 **16**	Nov 12 **17**	Nov 13 **18**	Nov 14 **19**	Nov 15 **20**	Nov 16 **21**
Nov 17 **22**	Nov 18 **23**	Nov 19 **24**	Nov 20 **25**	Nov 21 **26**	Nov 22 **27**	Nov 23 **28**
Nov 24 **29**	Nov 25 **30**					

Sunday:
Monday:
Tuesday:
Wednesday:
Thursday:
Friday:
Saturday:

Sunday:
Monday:
Tuesday:
Wednesday:
Thursday:
Friday:
Saturday:

Sunday:
Monday:
Tuesday:
Wednesday:
Thursday:
Friday:

Saturday:

Sunday:

Monday:

Tuesday:

Wednesday:

Thursday:

Friday:

Saturday:

Sunday:	
Monday:	
Tuesday:	
Wednesday:	
Thursday:	
Friday:	
Saturday:	

2019: 9th Moon (Nov 26 - Dec 24)

Sun	Mon	Tues	Wed	Thurs	Fri	Sat
		Nov 26 1	Nov 27 2	Nov 28 3	Nov 29 4	Nov 30 5
Dec 1 6	Dec 2 7	Dec 3 8	Dec 4 9	Dec 5 10	Dec 6 11	Dec 7 12
Dec 8 13	Dec 9 14	Dec 10 15	Dec 11 16	Dec 12 17	Dec 13 18	Dec 14 19
Dec 15 20	Dec 16 21	Dec 17 22	Dec 18 23	Dec 19 24	Dec 20 25	Dec 21 26
Dec 22 27	Dec 23 28	Dec 24 29				

Sunday:

Monday:

Tuesday:

Wednesday:

Thursday:

Friday:

Saturday:

Sunday:

Monday:

Tuesday:

Wednesday:

Thursday:

Friday:

Saturday:

RHYTHM TO THE UNIVERSE

Sunday:

Monday:

Tuesday:

Wednesday:

Thursday:

Friday:

Saturday:

Sunday:
Monday:
Tuesday:
Wednesday:
Thursday:
Friday:

Saturday:

Sunday:
Monday:
Tuesday:
Wednesday:
Thursday:
Friday:
Saturday:

RHYTHM TO THE UNIVERSE

2019-2020: 10th Moon (Dec 25 - Jan 23)

Sun	Mon	Tues	Wed	Thurs	Fri	Sat
			Dec 25 1	Dec 26 2	Dec 27 3	Dec 28 4
Dec 29 5	Dec 30 6	Dec 31 7	Jan 1 8	Jan 2 9	Jan 3 10	Jan 4 11
Jan 5 12	Jan 6 13	Jan 7 14	Jan 8 15	Jan 9 16	Jan 10 17	Jan 11 18
Jan 12 19	Jan 13 20	Jan 14 21	Jan 15 22	Jan 16 23	Jan 17 24	Jan 18 25
Jan 19 26	Jan 20 27	Jan 21 28	Jan 22 29	Jan 23 30		

Sunday:	
Monday:	
Tuesday:	
Wednesday:	
Thursday:	
Friday:	
Saturday:	

Sunday:

Monday:

Tuesday:

Wednesday:

Thursday:

Friday:

Saturday:

Sunday:

Monday:

Tuesday:

Wednesday:

Thursday:

Friday:

Saturday:

Sunday:

Monday:

Tuesday:

Wednesday:

Thursday:

Friday:

Saturday:

Sunday:

Monday:

Tuesday:

Wednesday:

Thursday:

Friday:

Saturday:

2020: **11th Moon** (Jan 24 - Feb 22)

Sun	Mon	Tues	Wed	Thurs	Fri	Sat
					Jan 24 1	Jan 25 2
Jan 26 3	Jan 27 4	Jan 28 5	Jan 29 6	Jan 30 7	Jan 31 8	Feb 1 9
Feb 2 10	Feb 3 11	Feb 4 12	Feb 5 13	Feb 6 14	Feb 7 15	Feb 8 16
Feb 9 17	Feb 10 18	Feb 11 19	Feb 12 20	Feb 13 21	Feb 14 22	Feb 15 23
Feb 16 24	Feb 17 25	Feb 18 26	Feb 19 27	Feb 20 28	Feb 21 29	Feb 22 30

Sunday:

Monday:

Tuesday:

Wednesday:

Thursday:

Friday:

Saturday:

Sunday:

Monday:

Tuesday:

Wednesday:

Thursday:

Friday:

Saturday:

RHYTHM TO THE UNIVERSE

Sunday:

Monday:

Tuesday:

Wednesday:

Thursday:

Friday:

Saturday:

Sunday:

Monday:

Tuesday:

Wednesday:

Thursday:

Friday:

Saturday:

Sunday:

Monday:

Tuesday:

Wednesday:

Thursday:

Friday:

Saturday:

RHYTHM TO THE UNIVERSE

2020: 12th Moon (Feb 23 - March 23)

Sun	Mon	Tues	Wed	Thurs	Fri	Sat
Feb 23 1	Feb 24 2	Feb 25 3	Feb 26 4	Feb 27 5	Feb 28 6	Feb 29 7
March 1 8	March 2 9	March 3 10	March 4 11	March 5 12	March 6 13	March 7 14
March 8 15	March 9 16	March 10 17	March 11 18	March 12 19	March 13 20	March 14 21
March 15 22	March 16 23	March 17 24	March 18 25	March 19 26	March 20 27	March 21 28
March 22 29	March 23 30					

Sunday:

Monday:

Tuesday:

Wednesday:

Thursday:

Friday:

Saturday:

Sunday:

Monday:

Tuesday:

Wednesday:

Thursday:

Friday:

Saturday:

Sunday:

Monday:

Tuesday:

Wednesday:

Thursday:

Friday:

Saturday:

Sunday:

Monday:

Tuesday:

Wednesday:

Thursday:

Friday:

Saturday:

Sunday:

Monday:

Tuesday:

Wednesday:

Thursday:

Friday:

Saturday:

This journal is the beginning of realizing that we have been led off-course by our man-made calendar and Holy Days! We follow a calendar that goes against the natural rhythm of the Universe! We are off course, it's no wonder people feel like they must find themselves!

Sensitive people, those in the arts and thinkers are thrown off by this, but they maybe can't seem to figure out what is wrong. We are being lied to about time. Everything on Earth follows this calendar – except for man.

We must respect the tribe of Judah for preserving it for all humanity. The Maccabees, Nehemiah, Issiah, Elijah, as many of the prophets found themselves the only one, or greatly outnumbered; and they fought, and they beseeched the Almighty, on the behalf of many who missed the mark, and even turned away. But, the Almighty, in eternal wisdom, blessed us and mighty in lovingkindness to all; but we know wickedness will not stand in high places in our eternal Kingdom come.

To the Messianic Jews, may they lead us into the coming new Kingdom here on Earth. And may we remember this message Jesus spoke in Luke. Who are my brothers and sisters? They are those who hear the word of God and obey it...

And to those who will stand like pillars in the Kingdom come, on Earth as it is in Heaven. When, here the Earth, will be as full of knowledge of the Lord, as the seas are full of water. Halleluiah!

RHYTHM TO THE UNIVERSE

GW Kushner

www.ingramcontent.com/pod-product-compliance
Lightning Source LLC
Chambersburg PA
CBHW031437210526
45464CB00005B/2236